YOUR KNOWLEDGE HAS VALUE

- We will publish your bachelor's and master's thesis, essays and papers

- Your own eBook and book - sold worldwide in all relevant shops

- Earn money with each sale

Upload your text at www.GRIN.com and publish for free

Alternative vehicle propulsion system with significantly reduced emission values based on cold liquefied air

Tivadar Menyhart

Bibliographic information published by the German National Library:

The German National Library lists this publication in the National Bibliography; detailed bibliographic data are available on the Internet at http://dnb.dnb.de.

ISBN: 9783346340573
This book is also available as an ebook.

© GRIN Publishing GmbH
Nymphenburger Straße 86
80636 München

All rights reserved

Print and binding: Books on Demand GmbH, Norderstedt, Germany
Printed on acid-free paper from responsible sources.

The present work has been carefully prepared. Nevertheless, authors and publishers do not incur liability for the correctness of information, notes, links and advice as well as any printing errors.

GRIN web shop: https://www.grin.com/document/992999

Alternative vehicle propulsion system with significantly reduced emission values based on cold liquefied air

Author: Dipl.-Ing. (FH) Tivadar Menyhart *

This document describes the development and calculation results of an efficiency-enhanced internal combustion machine (hereinafter referred to as BlueXTurb) with very low emission and consumption values, which can serve as an alternative vehicle power unit **. The conventional Joule-Brayton cycle is greatly improved by drastically lowering the compressor outlet temperature and increasing the turbine inlet pressure. The thermodynamic efficiency of the machine is increased by increasing the total pressure ratio to over 500:1. This is made possible by supplying the oxidizer in cold liquid form (approx. -190 ° C) with the greatest possible density and the least compressibility. The *turbo-electric power unit concept* promises the greatest advantages for a motor vehicle, i.e. a turbine that directly drives a (high-frequency) generator and can thus be kept at the optimal operating point. This means that electricity is available in a current/voltage conversion - with intermediate storage if necessary. The weight of the energy carrier - cold liquefied air and a small amount of liquid or gaseous fuel - including the vessel can be reduced by at least half compared to the traction battery of an electric vehicle of the same range. Gasoline, natural gas, ethanol, hydrogen, e-fuels or a gas mixture such as hythane can serve as a fuel. Compared to a modern, conventional Otto engine, the CO2 emissions of a car with a gasoline BlueXTurb power unit can be reduced by around 60-70% under realistic conditions. An H2-BlueXTurb enables a cost-effective and compact CO2-free propulsion system. The concept is able to meet the European CO2 emission limits, both as a car power unit and as a truck propulsion system by 2030.

Keywords:

BEV..........................	battery electric vehicle
BlueXTurb..................	expansion turbine with rotor, without a compressor, operated with a (cold) liquid oxidizer and a pressure ratio of at least $\Pi = 500$
dewar........................	vacuum-insulated vessel with thermal insulation
cold liquefied gas............	gases which boiling point is far below the ambient temperature; liquid air e.g. has a boiling point of approx. -190 ° C
LAIR.........................	liquid air
LOX..........................	liquid oxygen
oxidator.............	a chemical compounds or mixture that can oxidize other substances, i.e. is capable of releasing oxygen atoms.
turbo-electric drive..............	power unit comprising of a BlueXTurb, an electric generator and electric drive motor(s)

Symbols:

H_{low}	=	lower heating value [MJ/kg]	W	=	work [kJ]
Air_{St}	=	stoichiometric air requirement [-]	W_{eff}	=	effective work [kJ]
m	=	mass [kg]	w_t	=	specific technical work [kJ/kg]
p	=	pressure [Pa]	ε	=	compression ratio [-]
p_{final}	=	final pressure after compression [Pa]	κ	=	isentropic exponent [-]
p_{in}	=	intake pressure [Pa]	η	=	efficiency [-]
Q	=	heat quantity [kJ]	λ	=	combustion-air ratio [-]
Q_{suppl}	=	supplied heat [kJ]	η_{cc}	=	combustion chamber efficiency[-]
Q_{diss}	=	dissipated heat [kJ]	η_{inj}	=	injection pump efficiency [-]
V	=	volume [cm³]	η_{turb}	=	turbine efficiency [-]
V_d	=	displaced volume [cm³]	Π	=	pressure ratio [-]
V_c	=	compression volume [cm³]			

** German patent was granted on February 26th, 2020, international PCT procedure is in process (as of November 2020)

Table of contents

1. State of the art / problem definition

Legislation is setting ever stricter environmental requirements with regard to vehicle emissions. The EU plans to lower the CO2 emission limit for cars by 37.5% by 2030, and 30% are targeted in the commercial vehicle sector [01]. Conventional internal combustion engines will no longer be able to meet these requirements in the near future.

In spite of their low CO2 emissions, alternative drives have come under heavy public criticism. The fuel cell u. a. is very costly due to the platinum catalyst required (around 30-40 g per vehicle). The platinum price is likely to continue to rise as demand increases [02]. The achievable efficiency of a mobile unit is estimated at a maximum of 60%. The required hydrogen can only be produced artificially with a high expenditure of energy and is stored in expensive high-pressure containers - typically in CFRP containers with a wall thickness of a few centimeters, which today can only be produced using the complex, very expensive winding process. Typically, hydrogen gas is stored in a Type IV tank at a pressure of 700 bar. Country-specific safety regulations require a burst pressure of 1200-1575 bar though [03]. An H2 pressure tank takes up a lot of space due to the extremely low volumetric energy density of hydrogen. In a 150 L tank, only about 5.6 kg of hydrogen can be stored under 700 bar pressure, which covers a range of about 550 km at a consumption of 1 kg H2 / 100 km. On average, the retail price of an H2 fuel cell car is about twice as high as the retail price of a gasoline car [04], [05], [06].

Electromobility has been favored as a technology of the future for a number of years, but has not been able to establish itself across the board despite government funding. Long charging times, a lack of infrastructure, short ranges, especially when outside temperatures are low, are disruptive in everyday use. Heat losses due to fast charging, which deplete the overall well-to-wheel efficiency, or self-discharge when stationary are often not included in the overall efficiency analysis. The environmental pollution caused by the use of heavy metals, rare earths and solvents in large quantities weighs heavily. The mining of raw materials in often less developed countries under precarious circumstances, the increased energy consumption in the production, the disposal after wear and tear, which can be reached after 4-5 years, cause serious problems of the rechargeable battery technology [07]. Overall, a complete battery pack for a car weighs a few hundred kilograms. An electrically powered, fully loaded semi-trailer truck would require at least eight tons of batteries, which would require at least one megawatt of electrical charging power in a fast charging cycle. For such charging capacities on a broad scale, the power grid would have to be expanded first and supplied with sufficient energy from a larger number of power plants. The great technical breakthrough in accumulator technology has been hopefully expected for years in vain [08].

The thermal efficiency of internal combustion engines is largely determined by the compression or pressure ratio. The greater the pressure ratio, the higher the thermal efficiency (see **Figure 1**). That applies e.g. on the Diesel cycle, the Otto cycle, the Seiliger cycle, the Brayton cycle and the Joule cycle. For this reason, a higher compression or pressure ratio is aimed for in all of the thermodynamic cycle processes mentioned. The following shows how the engine pressure ratio can be increased in an expanded Joule-Brayton cycle - based on cold liquid air - in an expansion turbine.

In practice, compression-, expansion- and pressure ratios are technically linked to one another and cannot easily be separated from one another. A stronger compression causes a higher compressor outlet temperature, which in turn leads to a higher combustion temperature. However, this is limited by the thermal stability of the machine and cannot be chosen arbitrarily high. In reciprocating piston engines, the compression / expansion ratio is also limited by the knock limit of the fuel used. In the development of new generations of machines, an increase in the pressure ratio is always aimed for in the interests of higher efficiency, which leads to a massive technical effort [09]. As a consequence, the constructions must be able to withstand higher temperatures and pressures over the long term, which cannot be easily implemented. The interests of the thermodynamicist and the designer are diametrically opposed [10], [11].

$$\varepsilon = 1 + Vd / Vc \qquad\qquad\qquad\qquad \text{Eq. 1}$$

$$\Pi = p_{final} / p_{in} \qquad\qquad\qquad\qquad \text{Eq. 2}$$

$$\Pi = \varepsilon^{\,\kappa} \qquad\qquad\qquad\qquad \text{Eq. 3}$$

$$\eta_{Joule} = 1-(1/\Pi)^{\wedge}(\kappa-1)/\,\kappa \qquad\qquad\qquad\qquad \text{Eq. 4}$$

Table 1 – List of achievable pressure ratios and efficiencies in the best point; *Otto engine working according to the Atkinson cycle; **stationary gas turbine at sea level in continuous operation

	otto engine*	small diesel engine	gas turbine**
max. pressure ratio Π	(converted approx. 34:1)	(converted approx. 60:1)	33:1
max. compression ratio ε	17:1	23:1	(converted approx. 15:1)
max. efficiency η	38%	42%	42%

2. Thermodynamic basics

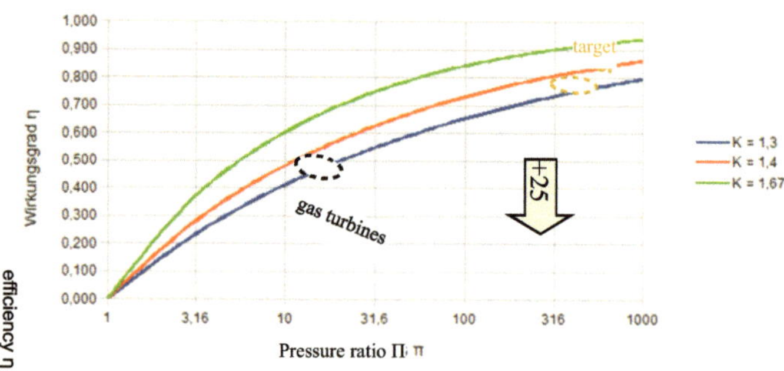

Figure 1 – Relationship between pressure ratio Π and theoretical efficiency η in the Joule cycle – taken from Eq. 4

What was sought was a cycle with the greatest possible pressure ratio, minimal compression work, and a low compressor outlet temperature (**point 2**), as well as the lowest possible exhaust gas temperature (**point 4**). The Joule / Brayton cycle serves as the basis. The area enclosed by the process (**1-2-3-4-1**) must be as large as possible and be as close as possible to the 0K line, then the ideal thermal efficiency is also maximal.

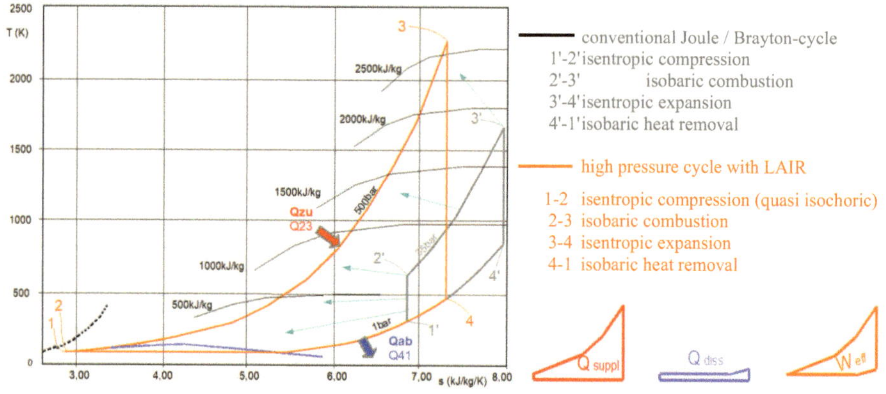

Figure 2 – conventional Joule-Brayton-cycle and high pressure cycle with LAIR in the T-S diagram, reversible changes of state

From Figure 2 it can be seen that the useful work (area **W eff** within **1-2-3-4-1**) is very large in relation to the amount of heat dissipated / anergy (**area Q diss**). The ideal (reversible) thermal efficiency is approx. 80% (according to Eq. 7). The specific useful work is about 4 times as large as in a comparable Joule / Brayton cycle. The same mass throughput results in four times as much useful work. Taken together with the very high density level, this indicates that a machine that is operated in the high pressure cycle process described achieves a much higher volumetric power density and mass power density than a modern gas turbine.

Explanation of the changes in state (reversible):

1 – 2 Isentropic compression

The pressure of the cold liquid oxidizer is increased from p1 = 1 bar to p2 = 500 bar by a liquid pump

The temperature is increased from T1 = 80 K to T2 = 91 K.

Supply of the specific compression work wt12 = 60 kJ / kg

The specific entropy remains constant

The specific volume decreases from V1=1,15 x 10^{-3} m³/kg to V2= 1,08 x 10^{-3} m³/kg

2 – 3 Isobaric combustion

The combustion takes place continuously in a flame zone

No pressure change

The temperature rises from T2 = 91K to T3 = 2250K

Heat supply q23 = 2750 kJ / kg

The specific entropy increases from s2 = 2.8 kJ / kg / K to s3 = 7.3 kJ / kg / K

The specific volume increases from V2 = 1,08 x 10^{-3} m³/kg to V3 = 1,35 x 10^{-2} m³/kg

3 – 4 Isentropic expansion

Expansion in a turbine

The pressure relaxes from p3 = 500 bar to p4 = 1bar

The temperature drops from T3 = 2250 K to T4 = 466 K

Withdrawal of turbine work wt34 = -2208 kJ / kg

The specific entropy remains constant

The specific volume increases from V3=1,35 x 10^{-2} m³/kg to V4= 1,38 m³/kg

4 – 1 Isobaric heat removal

The gas is released into the environment

Removal of specific heat q41 = -608 kJ / kg

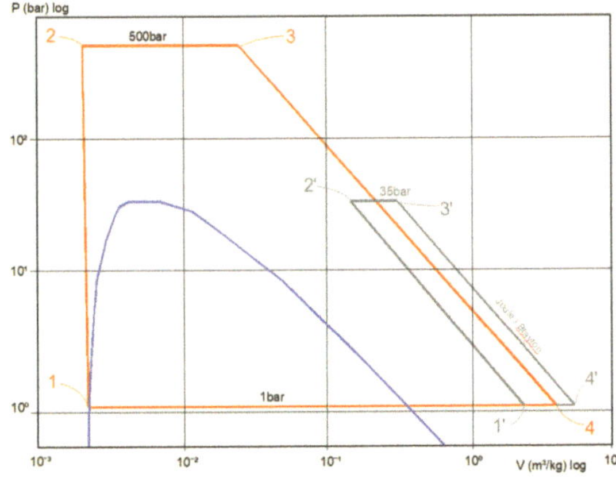

Figure 3 - conventional Joule-Brayton cycle and high pressure cycle with LAIR in the P-V diagram, reversible changes of state

The oxidizer (preferably LAIR) is supplied exclusively in the cold liquefied state. This eliminates the compression cycle. The liquid oxidizer is brought to the injection pressure below its critical temperature - quasi isochoric (i.e. almost no change in density / no volume change work) - and injected into the combustion chamber. Under real conditions, the injection pressure must naturally be somewhat higher than the combustion chamber pressure in order to overcome it. The isentropic pressure increase of cold liquefied air from 1 to 500 bar causes a temperature increase of approx. 11 Kelvin. The fluid density increases by about 7%, since liquids are also very slightly compressible under real conditions.

All oxygen carriers that can be stored in a cold liquefied state or in a liquid state at standard pressure are suitable as oxidizers. This includes e.g. Liquid oxygen (LOX), or oxygen-enriched, cold liquefied air. Cold liquefied air (LAIR) with natural oxygen, nitrogen and argon components is particularly suitable, as it is not toxic and does not pose a direct risk to the environment if it escapes in a controlled manner. In addition, it can be obtained anywhere from the ambient air and produced inexpensively [12]. Cold liquefied air costs around 18.4 € / t, or 1.6 ct / l to manufacture. The energy consumption is around 1021 kJ / kg or 247 Wh / l in the Heylandt process [13], [14]. Conservatively estimated, the production should require an average of approx. 300 Wh / L. Large plants such as the Cantarell Field Plant in Mexico, which produces tens of thousands of tons of liquid nitrogen every day, could probably be operated using even less energy. In comparison, the energy content of gasoline is approx. 9 kWh / L, or 32 MJ / L. The energy expenditure for the production of one liter of cold liquefied air corresponds to a gasoline equivalent of approx. 33 ml.

A major advantage of using liquid air instead of liquid oxygen is the greater mass throughput when a turbine is used (Eq. 5). Higher mass throughput results in more useful energy. 1 kg of gasoline requires around 14.7 kg of (liquid) air for stoichiometric combustion. When gasoline is burned with pure oxygen (O2), the ratio is around 1 kg gasoline to 3.5 kg oxygen. Breathing air consists of 78% by volume of nitrogen, which plays a far subordinate role / ideally no role at all during combustion. The inert nitrogen is also heated during combustion, changes its physical state from 'liquid' to 'supercritical gaseous' and thereby performs an enormous volume change work that is proportional to the increase in temperature. In terms of the highest possible energy density and thus the greatest possible range in the vehicle, the liquid air must reach the highest possible temperature before being introduced into the expansion unit and be expanded under the highest possible pressure ratio.

$$W_{eff} = wt_{34} \times m - wt_{12} \times m \qquad\qquad \text{Eq. 5}$$

$$Q_{suppl} = q_{H\,low} \times m \qquad\qquad \text{Eq. 6}$$

$$\eta_{eff} = W_{eff} / Q_{supp} \qquad\qquad \text{Eq. 7}$$

$$q_{cc} = q_{H\,low} \times \eta_{cc} = Hu \times \eta_{cc} / (\lambda \times Air_{St}) \qquad\qquad \text{Eq. 8}$$

$$\eta_{eff\,real} = 1 - (q_{Abgas} / q_{cc}) \qquad\qquad \text{Eq. 9}$$

or: $$\eta_{eff\,real} = (\Delta h_{turb} - \Delta h_{inj\,pump}) / q_{H\,low} \qquad\qquad \text{Eq. 10}$$

or: $$\eta_{eff\,real} = (wt_{34} \times \eta_{turb} - wt_{12} / \eta_{inj\,pump}) / (q_{H\,low} \times \eta_{cc}) \qquad\qquad \text{Eq. 11}$$

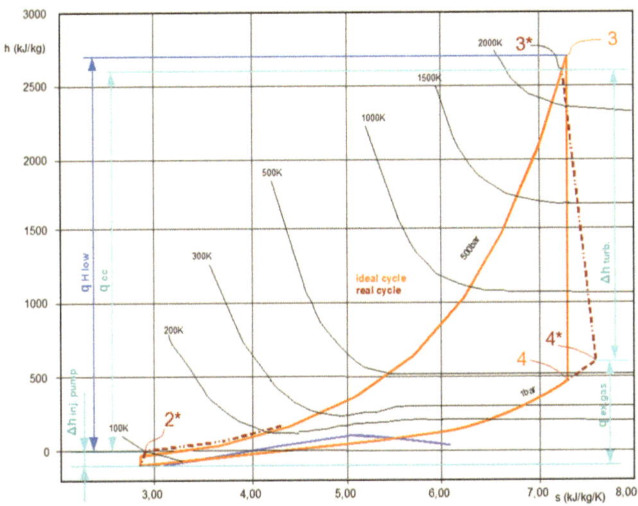

Figure 4 – high pressure cycle with LAIR cycle in the h-s diagram, ideal and real (irreversible)

From Figure 4 it can be seen that Δh inj. pump (real) is approx. 70 kJ / kg and Δh turbine (real) is approx. 2000 kJ / kg. Combustion chamber losses are already taken into account. This means that only around 3.5% of the turbine power is required to drive the injection pump. In a classic gas turbine, the compressor can require over 50% of the turbine output. Compared with the classic cycle processes (Otto approx. 1300 K, diesel approx. 1050 K, each under full load) the real exhaust gas temperature is a few hundred Kelvin (**point 4 ***, approx. 600 K) lower.

3. Thermal machine concept

Supersonic impulse turbines can process very high pressure ratios in one turbine stage. This type of turbine is very compact, has a small number of moving parts, the friction losses are low and there are no pulsation losses, as in a reciprocating engine. Therefore, the choice fell on this type of turbine. The enormous pressure ratio is achieved by connecting several supersonic turbine stages [15] in series or from a supersonic stage and one or more subsequent reaction stages. Single-stage supersonic turbines (impulse turbine) are used today e.g. in the ORC cycle as a turbo expander [16]. In the past, individual turbine stages were designed with a pressure ratio of up to $\Pi = 200$ [17]. In order to achieve such a ratio with high efficiency, the flow is accelerated to several times the local speed of sound in a Laval nozzle before entering the turbine. The total pressure ratio results from the product of the individual stage pressure ratios. The isentropic turbine efficiency η turb - taking into account aerodynamic losses, leakage losses, friction and bearing friction - can exceed 80% with an appropriate design [18], [19].

$$\Pi_{total} = \Pi_1 \times \Pi_2 \times \Pi_3 \dots \dots \times \Pi_n \qquad\qquad\qquad \text{Eq. 12}$$

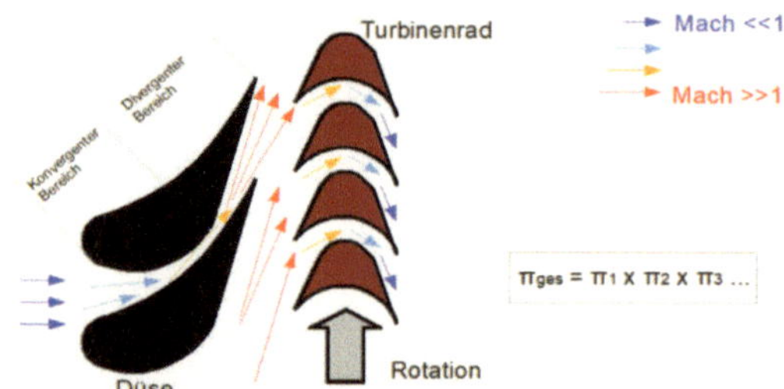

Figure 5 – Turbine stage of an axial supersonic turbine: stator - rotor

Another advantage of the high pressure LAIR Joule-Brayton cycle is that, in contrast to reciprocating piston engines (diesel, Otto cycle), there are no push-in and push-out losses due to the gas exchange. Overall, relatively low losses are to be expected in such a turbine system, which results from the injection pump efficiency η inj pump, required drive energy of the injection pump (s) Δh inj pump, combustion chamber efficiency η cc and turbine efficiency η turb (see Eq. 11).

The oxidizer releases its stored pressure energy after combustion with the fuel in the combustion chamber. The change in the physical state from liquid to gaseous happens after both have reached the combustion chamber. If the oxidizer is kept in the liquid state of aggregation until it enters the combustion chamber, i.e. below the critical temperature, the flow rate in the supply line remains very low - unlike in systems with large-area heat exchangers that require heat energy, e.g. taken from the environment. This means that throttling and flow losses remain low. The work required to insert it (liquid gas) into the combustion chamber is extremely small due to the low specific volume. The negative effects of the cryogenic temperatures, such as icing, condensation of air and oxygen enrichment, can be kept to a minimum by very short supply lines between the cryogen container and the turbine.

The greater the temperature difference reached, the more volume change energy can be released. A tremendous amount of volume change work is released. The density ratio of the combustion products (gasoline + LAIR) before and after combustion and expansion is around 1800: 1. In a stoichiometrically operated gasoline engine, the ratio is only 170: 1. To illustrate: Cold liquefied air would experience a pressure increase to approx. 10,000 bar in a closed container (initial pressure 1 bar) with isochoric heating from 80 K to 2000 K. The controlled pressure reduction in a dewar container is therefore essential.

The stoichiometric combustion temperature is about 2000-2400 K, depending on the fuel used, if LAIR is used as an oxidizer. If oxygen (LOX) is used, the combustion temperatures are might be above 3000 K. The fuel can be supplied in liquid or gaseous form. In adiabatic compression, however, gaseous fuels require more compression energy than liquids, which are almost incompressible. Since relatively little fuel is required in relation to the oxidizer, the losses are relatively low. The combustion pressure is 500 bar or higher. From a technical point of view, a chamber pressure of 1000 bar would be conceivable if one were to orientate oneself on modern common rail injection systems which today reach up to 3000 bar [20]. There peak load through pulsation represents a higher load than the static pressure alone. The combustion chamber of a 100-120 kW BlueXTurb is only a few cubic centimeters in size. The specified high combustion pressure and the high combustion temperature only occur in the combustion chamber and in the first guide vane or nozzle stage of the turbine. Behind this, both the temperature and the pressure drop suddenly. The high temperature and pressure load is therefore only a challenge in a very small area of the machine.

The combustion chamber of a BlueXTurb is very similar to the combustion chamber of a liquid rocket. Rocket combustion chambers achieve efficiencies of 95%-99.5% [21]. This is due to the good distribution, mixing and turbulence of the combustion gases. The injection plate of a rocket engine consists of several perforated discs connected to one another. At the same time, it represents the cover plate of the cylindrical combustion chamber and injects fuel and oxidizer over a large area through many small holes over its entire surface.

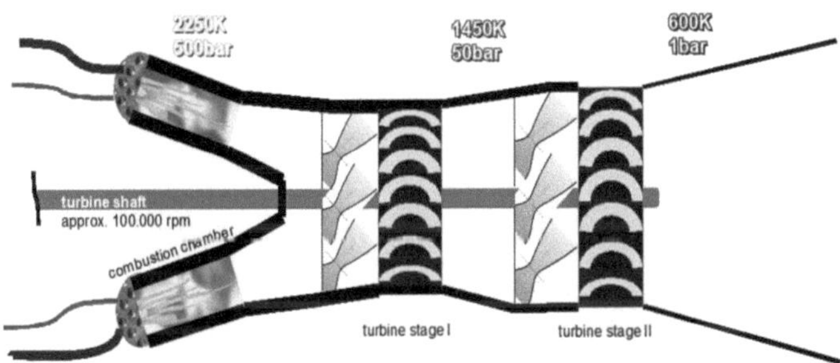

Figure 6 – Schematic structure of a BlueXTurb with two axial supersonic turbine stages

4. Problem of the thermal load in the turbine

The turbine inlet temperature of up to 2400 K described is around 500 K too high for a gas turbine according to the current state of the art. The metal materials used cannot withstand such high temperatures over the long term. Up to 1900 K are common in large gas turbines with curtain or film cooling. The high combustion temperature is also seen as an opportunity to improve combustion and thereby reduce exhaust emissions. The desired flameless combustion has been researched and can be achieved through the following factors:

- High combustion temperature, above the auto-ignition temperature
- Strong turbulence in the combustion chamber
- High exhaust gas recirculation rate thanks to a special burner shape, thereby lowering the oxygen concentration in the highest temperature area

The formation of nitrogen oxide (NOX), for example, can be reduced by over 70% compared to normal combustion. The combustion is cleaner and more complete [22].

The following four possible solutions are proposed to lower the turbine inlet temperature:

Version 1: The turbine is operated with an excess of oxidizer. Similar to the curtain or film cooling in a gas turbine, additional air is supplied to the combustion chamber walls, turbine stator and rotor. The air consumption compared to stoichiometric combustion is increased accordingly. The thermodynamic disadvantage of the lower turbine inlet temperature is roughly offset by the advantage of the increased mass throughput. Fuel consumption and CO2 emissions hardly change. The higher air consumption must, however, be covered by a larger liquid air tank for the same range. The increased nitrogen oxide formation can be minimized, for example, by a low-temperature SCR catalytic converter with urea injection, which is already active from 200 °C [23].

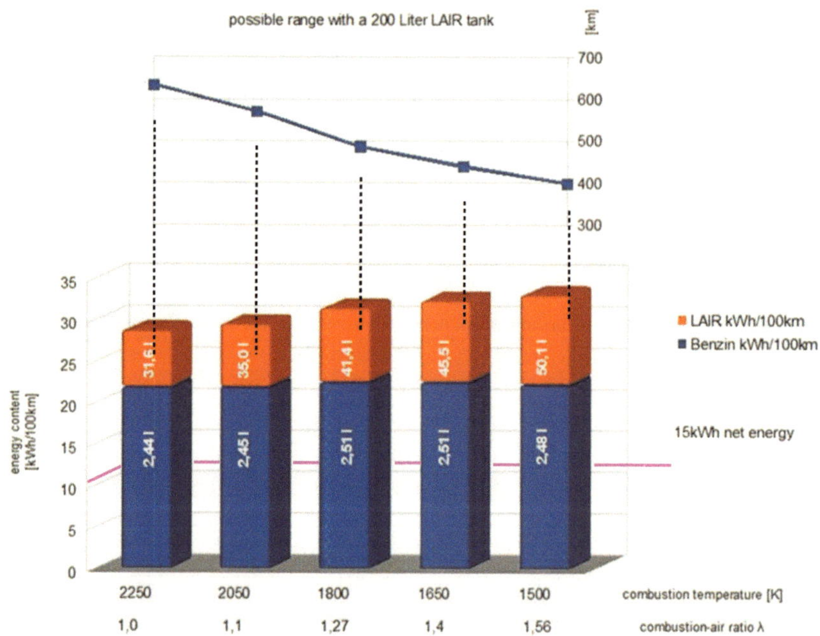

Figure 7: Effect of the combustion-air ratio on the combustion temperature, fuel consumption, air consumption and the range/milage of a passenger car equipped with a 200 liter liquid air tank - the basis is a 1.5 ton car with 15kWh / 100km energy requirement

Version 2: Additional water injection into the turbine with the aim of cooling (similar to version 1). The water required for this can be carried fresh or recovered from the exhaust gas. In every combustion process, a not inconsiderable amount of water is produced anyway. Example: If Gasoline-air is combusted in correct stoichiometric ratio the result is approx. 8% water vapor in the exhaust gas. Due to the inherently low exhaust gas temperature, condensed water can be recovered from the exhaust gas tract with the help of heat exchangers.

Version 3: The turbine is divided into a high pressure and low pressure area. After the isentropic expansion in the high-pressure part, the flue gas is directed past the outer wall of the combustion chamber or passed into a heat exchanger and temporarily heated there. The combustion chamber can also be designed in such a way that it protrudes into the area of the second turbine stage and emits heat there. The heat exchanger can be fed from the heating chamber or the stator(s). This leads to a lower turbine inlet temperature before the high pressure turbine and a higher turbine inlet temperature before the low pressure turbine. As a result of the reheating (compare Clausius-Rankine process), the overall efficiency only drops slightly. The pressure loss in the heat exchanger and the longer gas paths, however, have a negative effect.

Version 4: High temperature-resistant composite materials with a very low coefficient of thermal expansion are already used in rocket engines in the area of the nozzle neck or as heat protection tiles. For the next generation of gas turbine blades, a fatigue strength of over 2000 °C could be achieved in the coming years through the use of ultra-high temperature-resistant ceramic matrix composites (UHTCMC) [24]. These can be used in particularly high-temperature areas, such as the combustion chamber walls and the first turbine stage. The cost factor is low because the affected parts are much smaller than in a conventional gas turbine. The turbine runner of a 100-120 kW BlueXTurb is estimated to be a few centimeters in diameter. The 3D printing of such materials for use in turbines [25] is being tested. The design freedoms could be exhausted to a degree still unimagined today for high aerodynamic efficiency.

A combination of several versions is conceivable and is possibly the best solution.

5. Propulsion system concept and vehicle structure

A BlueXTurb can be coupled directly to an e-machine (e.g. a high-frequency generator) (Fig. 8). Compare 'turbo-electric propulsion' in locomotives and large ships. The generator - which in practice continuously achieves an efficiency of up to 99% [26] - can e.g. in start-up operation be reversed to an electric motor operating state and thus the turbine if necessary may be driven for a faster start-up. The electrical current generated in this way can be temporarily stored after voltage / frequency conversion in a small accumulator or in capacitor(s) or a combination of both before the electrical current is sent to the drive machine(s). This structure enables recuperation of braking energy. The current can be sent directly to the electric drive motor(s) without buffering. Voltage converters, power electronics and drive motors can be taken from a BEV (i.e. with identical degrees of efficiency). The turbine can thus be kept in one or more optimized operating point (s) for the highest possible efficiency. The control of the turbine, or regulation of the battery charge level, is carried out by an intelligent management system, which ideally includes navigation data in the regulation. This can e.g. with the help of topographical data, decide whether electrical energy is required or whether storage capacity should be kept ready for braking energy.

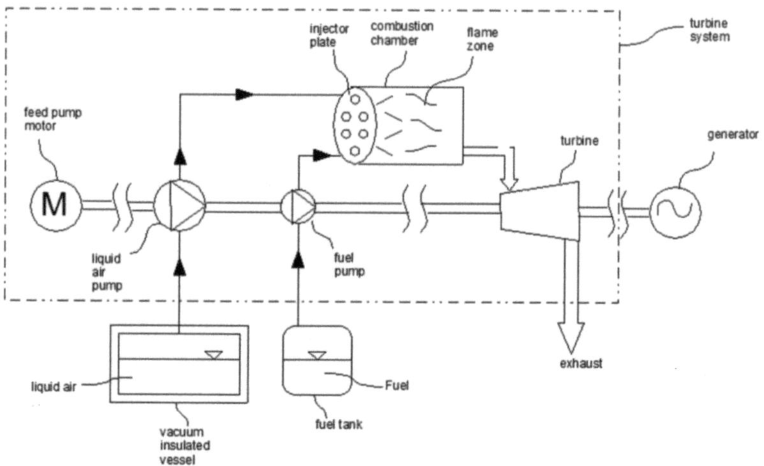

Figure 8 – Schematic structure of the turbine unit and the peripherals in the vehicle

Fresh gas volume and turbine speed are not related to each other, as is the case with the single-shaft gas turbine. The flow rate of the feed pump of a BlueXTurb can be regulated independently of the turbine speed. This enables a very high dynamic. The machine needs considerably less time to reach a desired load point than, for example, an exhaust gas turbocharger. In addition to that, the coupled generator can be polarized to support the electric motor operation during the start-up process. The BlueXTurb can be started up extremely quickly and it quickly provides the required electrical energy. Therefore, no such large vehicle battery / backup battery is necessary, such as in a range extender operation. The battery only needs to be able to bridge or support a few seconds, e.g. is accelerated on a slope. It mainly serves as a storage for recuperation energy during braking, for which a unit with 1 kWh capacity is completely sufficient. The weight is max. 30 kg including housing and cooling.

For comparison: The Li-Ion battery pack of a Formula 1 drive unit from the year 2018 including power electronics had to be between 20 and 25 kg. Capacity: 4 MJ = 1.1 kWh. Power max. 120 kW. This maximum output can thus be delivered for about 33 seconds [27]. Units with a higher energy density (kWh / kg) that are used in e-cars (BEV) do not achieve such high power densities (kW / kg) [28].

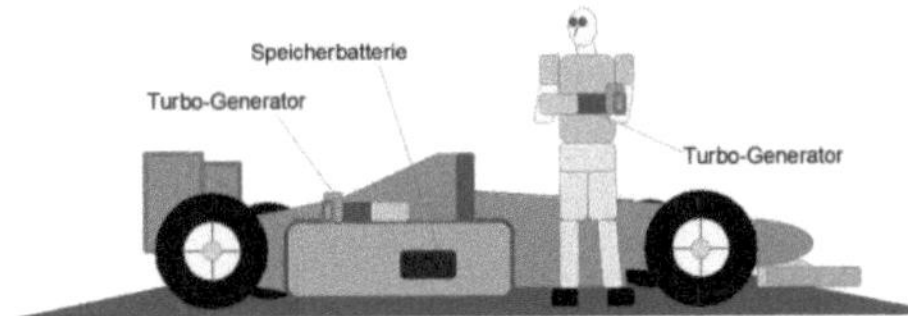

Figure 9:
Reference: MGU-H unit of an F1 engine „Motor-Generator-Unit Heat". The e-machine is driven by the turbocharger shaft with up to 120,000 rpm. Nominal power consumption: 50 kW.

motor-generator-turbine unit

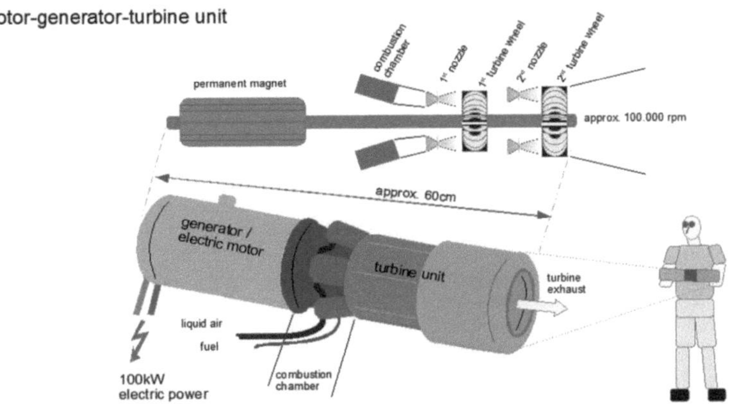

Figure 9 – Concept and proportions of a compact BlueXTurb turbine generator unit, passenger car size: approx. 100-120 kW power output

A 100-120 kW BlueXTurb achieves, due to the large specific useful work and the high fluid density, a very high power density, which is also considerably higher than that of a gas turbine. The weight is estimated to be less than 10kg. External dimensions and production costs are comparable to an exhaust gas turbocharger in today's passenger car. The BlueXTurb generator unit can be placed in a car where the transmission tunnel is today. The large vacuum-insulated container for the oxidizer can be placed in the front of the vehicle. 200 liters of liquid air (approx. 174 kg), combined with fuel (e.g. approx. 15 liters of gasoline, or 3.5 kg of hydrogen, or approx. 10 kg of natural gas) in a 1.5 t car (energy consumption approx. 15 KWh / 100km) enable up to 630 km range. The ratio between fuel and LAIR varies depending on the type of fuel and air ratio (see Fig. 7). Since manual transmission and combustion engine are not required, the total weight of a vehicle with BlueXTurb drive is roughly comparable to a vehicle with Otto drive, with a similar range.

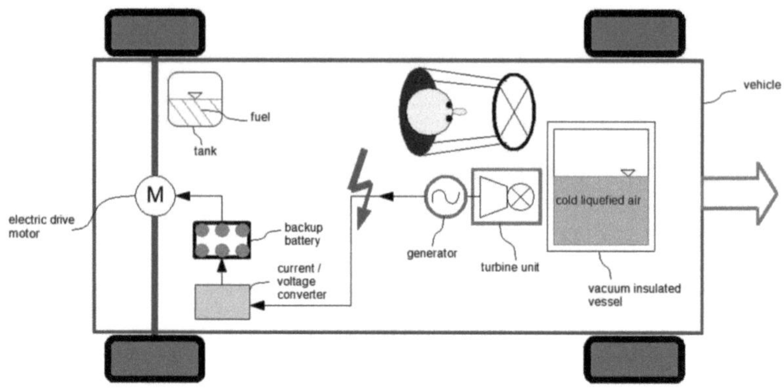

Figure 10: Schematic structure of a vehicle with a BlueXTurb generator unit

The high density state of the fluid is maintained by cold and not by pressure. To maintain the high density, no thick-walled, heavy pressure accumulator is necessary, but a cold-insulated container (synonyms: Dewar container, cryogenic vessel), which is constructed like a thermos in the broadest sense. The most efficient thermal insulation possible is essential so that the contents heat up as slowly as possible. Such containers are protected against radiant heat (mirrored surfaces), heat conduction and convection (small contact surfaces between the inner and outer walls, evacuated spaces). As early as the early 2000s, cryogenic containers for vehicles were able to store 120 liters of liquid hydrogen (boiling temperature 25 K / approx. -250 ° C) at 1-5 bar internal pressure for up to three days without the pressure having to be released [29]. The latency period of LAIR in an equally well insulated container is even longer, since the boiling temperature is significantly higher (83 K / approx. -190 ° C) and the density of the fluid is approx. ten times higher. As a result, the volume and surface of the corresponding liquid air container is considerably smaller. This ensures a much smaller flow of heat into the interior of the container. The larger the container, the less the negative influence of the heat input.

In a car with a kerb weight of 1500 kg, a container of around 200 liters is seen as an ideal compromise between installation space, weight, range and production costs. The weight of the container is estimated at 70-100 kg, plus 174 kg of content (LAIR). The battery unit of a comparable battery-electric vehicle with a similar range is more than twice as heavy and the production costs are estimated to be around 20-50 times as high [30]. The average fill volume of a LAIR tank is only 50% of the tank capacity between two fillings. In contrast, the weight of a battery does not depend on the state of charge. A combined container for liquid hydrogen with an insulating layer of liquid air with a latency period of up to 12 days has been successfully tested in the past. It consists of an internal container for liquid hydrogen with an external liquid air container wrapped around [31]. This technology could establish itself for the H2-BlueXTurb drive.

Liquid oxygen and nitrogen have different boiling temperatures. Therefore, with a controlled pressure release, oxygen would accumulate in the liquid air tank, which on the one hand leads to a falsification of the stoichiometric ratio and on the other has a strong fire-accelerating effect. The blow-off gas would contain a lot of nitrogen, which is particularly dangerous in closed rooms, underground garages, etc. The use of a device that prevents segregation or oxygen enrichment in the container is therefore very useful in terms of operational safety. A passive heat exchanger system cools the vaporized, but still cold, nitrogen-rich gas in the container back below the dew point with the help of liquid air, and the pressure drops again. Only evaporated air with unchanged composition emerges. Such systems have been used for years, for example in breathing air cylinders [32].

The simultaneous non-pressurized refueling of liquid air and fuel can be done in a matter of minutes using a combined, cold-insulated connection hose, which prevents the mix-up of operating substances. The greatest danger posed by cold liquefied air is frostbite / cold burn on contact. This must certainly be ruled out by taking appropriate precautions at the fueling system, crash safety in the vehicle, etc. Broken down vehicles or completely empty liquid air tanks can under certain protective measures - e.g. wearing cold-insulating gloves

– be refuelled by hand easily. Poorly insulated, portable cryogen containers made of aluminum with a capacity of a few liters are already available today, e.g. for the laboratory sector. Such containers could establish themselves as inexpensive, 'emergency canisters'.

6. Heating and air conditioning

Due to the high pressure ratio, the exhaust gases cool down to an unusually low turbine outlet temperature of around 300-650 ° C. This is a huge drop in relation to the high combustion temperature (up to 2130 ° C) (Fig. 4, **point 4 ***). As already described, the exhaust gas temperature depends on the stoichiometric ratio, the combustion pressure and the fuel and its typical combustion temperature. Therefore, compared to classic internal combustion engines, there are very low thermal losses. The residual heat in the exhaust gas can still be used to heat up the passenger compartment. In contrast to electric cars, no or only relatively little electrical energy is required for additional heating (e.g. electrical seat heating or window heating). In addition, there is no temperature control of the traction batteries or, due to the low battery mass, requires considerably less energy for temperature control. E-cars also have to be temperature controlled during the charging process and thus lose large amounts of energy.

The outside air supply or the circulating air in the interior can be enriched for cooling with supplied cold liquefied air from the vacuum-insulated container. This can be done by spraying the liquid air. Alternatively, targeted evaporation in an evaporation container can be used for cooling via a surface humidified with liquid air or a moistened membrane. As an alternative or in addition to the liquid supply, the amount of evaporation that is generated in the vacuum-insulated container anyway due to boil-off can be introduced into the passenger compartment in a metered manner. In both cases it makes sense to filter the gaseous air produced in this way, e.g. in a charcoal filter before being introduced into the interior. Such a system does not require any lossy, heavy high-pressure compressors or pumps. There is no need for a heavy, energy-intensive air conditioning compressor, refrigerant or heat exchanger. The air conditioning does not generate any waste heat and no additional local CO_2 emissions. All that is needed is a small feed pump. Under certain circumstances, the overpressure of the vacuum-insulated vessel alone is sufficient to convey and atomize the required cold-liquefied air.

The cooling requirement of a mid-range car in a stationary state is on average approx. 200-250 W. As a result, the liquid air consumption of a car increases by approx. 2.5 - 3 liters per hour through active air conditioning using liquid air. On the other hand, there is the nominal output of an average air conditioning compressor of approx. 3-6 kW, which has to be reserved on the hardware side for peak loads, but which is used considerably less in normal operation. Such a compressor is no longer required. Classically, the air conditioning only runs at full load when a parked vehicle that has become extremely hot in strong sunlight is cooled down to moderate temperatures in a short time.

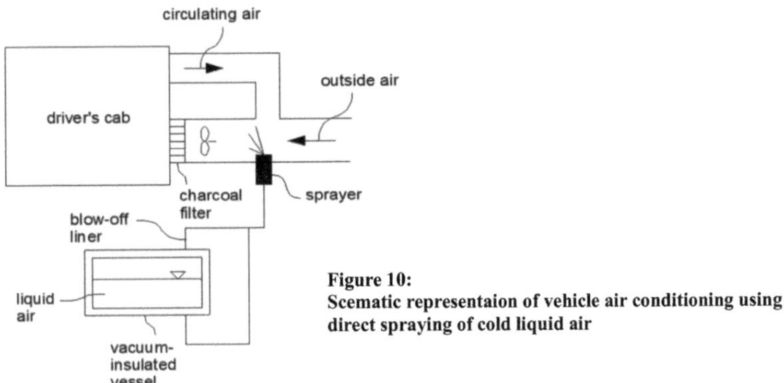

Figure 10:
Scematic representaion of vehicle air conditioning using direct spraying of cold liquid air

7. Fuel efficiency

Reaction Equations:

Gasoline (octane): $\quad$ $2C8H18 + 25O_2 + 94,05N_2 \quad \rightarrow \quad 18H_2O + 94,05N_2 + 16CO_2$

$\qquad$ 1 kg $\quad$ 3,505 kg $\quad$ 11,54 kg $\quad \rightarrow \quad$ 1,42 kg $\quad$ 11,54 kg $\quad$ 3,086 kg

Methane: $\quad$ $CH4 + 2O_2 + 7,524N_2 \quad \rightarrow \quad 2H_2O + 7,524N_2 + CO_2$

$\qquad$ 1 kg $\quad$ 3,99 kg $\quad$ 13,135 kg $\quad \rightarrow \quad$ 2,245 kg $\quad$ 13,135 kg $\quad$ 2,742 kg

Hydrogen: $\quad$ $2H_2 + O_2 + 3,762N_2 \quad \rightarrow \quad 2H_2O + 3,762N_2$

$\qquad$ 1 kg $\quad$ 7,92 kg $\quad$ 26,09 kg $\quad \rightarrow \quad$ 8,92 kg $\quad$ 26,09 kg

Simplified air composition is assumed in the reaction equations (mass fraction) - no argon and trace gases:

$$Air = 23,3\% \ O_2 + 76,7\% \ N_2$$

Table 2: Specific fuel consumption and efficiency at the optimal load point - the calculation basis for the efficiency is Eq. 11 ($\Pi = 500$; $\lambda=1$; $\eta_{cc} = 0,95$; $\eta_{turb} = 0,8$)

propulsion system	specific fuel consumption	total efficiency
passenger car Otto engine	230 g/kWh	38%
passenger car turbodiesel	200 g/kWh	42%
truck turbodiesel	190 g/kWh	45%
BlueXTurb gasoline	122 g/kWh	68%
fuel cell H2	60 g/kWh	50%
BlueXTurb H2	45 g/kWh	68%

Table 3: Calculated BlueXTurb turbine consumption and pollutant emissions of an approx. 1.5 ton passenger car with an electrical energy requirement of approx. 15 KWh / 100km - stoichiometric air ratio

type of fuel	fuel consumption	liquid air consumption	CO2 emissions (local)
gasoline	2,44 L/100km	31,6 L/100km	56 g/km
methane (natural gas)	1,63 kg/100km	32 L/100km	45 g/km
hydrogen (H2)	0,66 kg/100km	26 L/100km	------------

Table 4: Calculated BlueXTurb consumption and pollutant emissions of a fully loaded 40 ton truck with an electrical energy requirement of approx. 125 kWh / 100 km; Diesel consumption of a truck today: approx. 32-40 liters, approx. 1 kg CO2 / km; H2 consumption of a fuel cell truck: approx. 8-10 kg [33]

type of fuel	fuel consumption	liquid air consumption	CO2 emissions (local)
gasoline	20,3 L/100km	263 L/100km	470 g/km
methane (natural gas)	13,6 kg/100km	268 L/100km	374 g/km
hydrogen (H2)	5,56 kg/100km	217 L/100km	-----------

For a holistic assessment of the BlueXTurb unit as a car drive, various aspects were listed in a table (Table 5). The calculation is based on a medium-sized car with BlueXTurb weighing approx. 1500 kg empty and approx. 120 kW nominal power. The range for all vehicles is around 600 km. The consumption was determined according to the WLTP cycle and increased by 25% in order to obtain a more realistic, practical value. For reasons already mentioned, it can be assumed that a BEV with a comparable interior size and similar payload is around 400-600 kg heavier and, due to the higher weight, the slightly larger frontal area, and the increased electrical energy requirement (battery temperature control, electrical heating, air conditioning) has a specific energy consumption that is approx. 20% higher. The same applies to a fuel cell vehicle that is 200-300 kg heavier. Since e.g. the brake system, chassis components, crash structures have to be reinforced for an increased battery weight, a higher battery weight has a disproportionate effect. The vehicle with a 2.0 L diesel engine (approx. 150 kg engine weight + 90 kg automatic transmission) is likely to be as heavy as a BlueXTurb vehicle (approx. 1500 kg empty weight). The specified energy requirement includes the losses that occur in the buffer battery, the power electronics and the electric drive motors (including fixed gear ratio).

Table 5: Comparison of several types of drive for a medium-sized class car with comparable space and payload under different aspects*.

propulsion system	BlueXTurb gasoline @500 bar λ=1,0	BlueXTurb H2 @500 bar λ=1,0	E-car (BEV)	fuel cell (FCEV) 120kW	diesel engine 2,0 L 120kW
consumption / 100km [WLTP + 25%]	2,44 L gasoline + 31,6 L LAIR	0,66kg H2 + 25,8 L LAIR	18kWh Strom	1,1 kg H2	5,5 L diesel fuel
provided electr. energy upstream of the power electronics	15kWh / 100km	15kWh / 100km	18kWh / 100km	18kWh / 100km	---
CO2 emissions / car	56 g/km	0 g/km	0 g/km	0 g/km	145 g/km
CO2 emissions / power plant 474g CO2/kWh German average	37 g/km	149 g/km - H2 30 g/km - LAIR	85 g/km	248 g/km	0 g/km
CO2 emissions / sum	93 g/km	179 g/km	85 g/km	248 g/km	145 g/km
fuel cost	6,6 €/100km	9,2 €/100km	5,22...16,02 €/100km	11 €/100km	6,9 €/100km
vehicle weight / tank is full	1500kg	1550kg	2000kg	1850kg	1500kg
tank weight with content for approx. 600km range	2 kg gasoline tank 11 kg gasoline 166 kg LAIR 90 kg dewar vessel	66 kg H2 pressure tank 4 kg H2 gas 155 kg LAIR 90 kg dewar vessel	750 kg traction battery	110 kg H2 press.tank 6,6 kg H2 gas	4,5 kg diesel tank 24,75 kg diesel fual
manufacturing costs for the energy delivering machine	500€ BlueXTurb	500€ BlueXTurb	---	7500€ fuel cell	2000€ basic engine with auxiliaries
manufacturing costs for the energy container - 100km range each	250€ für 1kWh buffer battery	3300€ pressure tank type IV / 700bar	13750€ traction battery	5500€ pressure tank IV 250€ buffer battery	---
opportunities / challenges	turbine is very small, inexpensive, quick refueling possible Turbine has yet to be developed	no CO² emissions during combustion H² pressure vessel is expensive	locally emission-free Energy requirement f. manufacturing, recycling Charging times, -stations Charging losses	complete Fuel cell system is very expensive and relatively inefficient, 30-40g platinum required	required exhaust gas and CO² emissions techn. not solvable thermodynamic limits
additional expenditure on infrastructure	petrol stations with LAIR needed	petrol stations with H2 + LAIR needed	more charging points, electricity grid expansion	petrol stations with H2 needed	---

*** Electricity and fuel prices:**
Electricity generation in the federal mix 2018 – 474 g CO²/kWh – German Environment Agency; [34]; charging losses not taken into account
Electricity price at fast charging stations AutoBild, June 2019 „Öffentlicher Ladestrom teurer als Benzin" [35]

German average fuel price, as of February 2020: 1,25 €/L diesel fuel, 1,40 €/L premium gasoline
German average H2 price provided at gas stations, as of February 2020: 10 €/kg
Estimated price for liquid air at gas stations including tax: 0,10 €/L
propellant transport costs, charging losses, boil-off losses, self-discharge were not taken into account

Due to the low specific energy requirement of the BluexTurb, the combustion of natural gas or hydrogen could prevail in addition to gasoline, diesel fuel or artificially produced e-fuels. A biologically produced mixture of hydrogen and methane (hythane), which has a lower volumetric calorific value, but a higher density than pure hydrogen, would also be possible. Hythane typically consists of 8-32% by volume hydrogen.

Hydrogen can be stored as a gas under high pressure, which means a high energy requirement for pressurization. Heavy CFRP pressure vessels are required for typically 700 bar system pressure. Alternatively, hydrogen can be stored chemically in an organic carrier material (LOHC) without pressure [36], or in cold liquefied form (-253 ° C).

So far, fuel cell vehicles have not caught on mainly because of the high manufacturing costs. The pressure tank and the platinum requirement of the fuel cell play a large role in the production costs. The specific hydrogen consumption (g / kWh) of an H2-BlueXTurb drops by approx. 25% compared to the fuel cell. In all cases, a significantly smaller pressure vessel or H2-Dewar container would be required for the same range than in a vehicle with a fuel cell. This not only lowers production costs, but also benefits the overall vehicle weight and thus in turn lowers consumption.

The BlueXTurb is particularly suitable as a truck propulsion system, since the payload is roughly retained and fast refueling is possible without a special charging infrastructure and without further power grid expansion. The disproportionately high purchase price for a traction battery does not apply. With a 2000 liter dewar container for LAIR, the range with petrol BlueXturb is around 750 km. Although this is considerably less than the range of a truck with a diesel drive, the local CO2 emissions are roughly halved and quick refueling is possible regardless of the power grid. In addition, propulsion with methane gas (approx. 65% less CO2 emissions than diesel propulsion), in combination with a similar LAIR range (750 km). The H2-BlueXTurb drive enables CO2-free operation. For a range of 750 km, a total of 1200 L hydrogen pressure tanks in combination with one or more Dewar tanks with a total of 1630 L are required for LAIR. The hydrogen consumption is reduced by approx. 25-30% compared to a fuel cell drive (Table 4).

The BlueXTurb drive is also suitable for (city) buses, railway locomotives on routes where no electrical overhead line is (yet) available, or as a ship propulsion system.

8. Energy industry / infrastructure

In the future, the rationing of charging current will be inevitable [37] if the majority of vehicles rely on the electrical power grid as an energy source and this is not expanded. Therefore, in the future, electrical energy should be stored in large unpressurized tanks in the form of cold liquefied air in the event of overcapacities in power plants and stored for reconversion [38], whereby very large tanks can get by for up to 150 days without blow-off. The use of cold liquefied air as vehicle fuel is not planned today, but would make ecological sense. Liquefaction results in far lower losses than the compression work in an internal combustion engine or in a gas turbine. The liquefaction happens z. B. in the Claude process in a stationary air liquefaction plant. The raw material air is available free of charge everywhere on earth, is non-toxic, non-polluting and can be stored and transported inexpensively in vacuum-insulated containers similar to liquefied natural gas at -190 ° C. Storage is possible without geographical restrictions, for example in large sea tankers, or on the rails in tank wagons. A further expansion of the electrical power grid would not be necessary to the extent that grid-dependent electromobility requires.

In a power plant, renewable electricity can mainly be used at night in the cheap electricity phase for liquefaction, or if there is overproduction anyway. The resulting pressure energy is stored in the oxidizer (preferably in cold liquefied air), which is largely released again during combustion in the combustion chamber of the BlueXTurb. The stronger the warming, the more energy is released. It makes sense to consider the sum of the energy of the fuel and the oxidizer that is released during combustion.

$$W = p \times V \qquad\qquad \text{Eq. 13}$$

Table 6: Evaluation of the energy density of the energy carrier including the container, i.e. e.g. Gasoline + gasoline tank + LAIR + dewar container; The useful energy results from the efficiency at the best point from Table 2

propulsion system	BlueXTurb gasoline	BlueXTurb H2	E-drive (BEV)	fuel cell H2 (FCEV)	diesel engine
gross energy content with full tank and pressure vessel @700bar	0,49 kWh/kg	0,42 kWh/kg	0,15 kWh/kg	1,85 kWh/kg	1,84 kWh/kg
of which can be used at the optimal load point	0,33 kWh/kg	0,29 kWh/kg	0,14 kWh/kg	0,93 kWh/kg	0,78 kWh/kg

9. Conclusions

Excess energy from renewable energy sources (solar, wind, water power) could be stored in liquid air. Experimental systems for energy storage using liquid air are currently being tested in several countries. The energy stored in this way can also be used sensibly in the transport sector.

Air as a raw material is available free of charge anywhere on earth and can be stored and transported in a similar way to liquefied natural gas. It can be produced e.g. in sunny countries with the help of solar power, or near the coast with wind power. The production of 1 liter of liquid air today costs about 1.6 cents with 247 Wh energy consumption (Heylandt process).

A BlueXTurb drive unit has considerably less weight and is cheaper to manufacture than a diesel or gasoline engine. The manufacture of the turbine unit and the vacuum-insulated container for the liquid air therefore requires very little energy and resources. Environmentally harmful rare earths, lithium, cobalt, coltan etc. are required in much smaller quantities than in an electric car (BEV). A small backup battery, or capacitor, only needs to be able to support the turbine for a few seconds, or it serves as a store for braking energy. Aging or self-discharge of the buffer battery has fewer negative effects here. In addition, there is significantly less environmental damage during disposal or recycling.

Due to the high level of efficiency, which also exceeds a fuel cell, a CO_2-free drive with hydrogen - without the use of expensive platinum catalysts and with significantly smaller H_2 pressure tanks than in a mobile fuel cell - is finally affordable and technically feasible. The truck division, which is under pressure to reduce CO_2 emissions with the lowest possible loss of payload, could benefit most.

There are no hours of charging and there is no dependency on the power grid. Refueling is possible in minutes using a cold-insulated hose. Every petrol station that exists today can be retrofitted with a vacuum-insulated tank for cold liquefied air - even underground. The fuel and liquid air can be filled up at the same time via a combined (safety) filler neck.

Driver's cab air conditioning and cooling is carried out with the aid of the liquid air that is carried along, which is directly fed into the air supply in sprayed form. This increases the air consumption somewhat. Expensive, ineffective air conditioning compressors, heat exchangers and environmentally harmful refrigerants are not necessary. There is no waste heat and no local CO_2 emissions from using the air conditioning. Air conditioning is also possible when the vehicle is stationary without the propulsion system running. The residual heat from the approx. 300-650 ° C warm exhaust gas can be used for heating.

Since the propulsion system is largely made up of already known and mastered technologies, the technical risks are to be assessed as relatively low. The technology could be developed to pre-series maturity in a year in a company with the appropriate background - also considering the technical challenges. The greatest difficulty is undoubtedly the high turbine inlet temperature for the first turbine stage. The temperature reduction through internal cooling, i.e. through excess oxidizer or water injection, represents a promising approach for implementation with materials that are already available today, but worsens the efficiency by a few percentage points. A more detailed assessment of the possible scenarios is only possible through calculation, simulation and the creation of detailed concept models.

Reference

[01]   Christian Frahm, Strengere CO2-Grenzwerte: Jetzt gibt es kein Zurück mehr. Spiegel Online. 18. Dezember 2018
         (www.spiegel.de, 11. Januar 2019)

[02]   Frank Wiebe, Jakob Blume, Anleger hoffen auf Platin – dank der Brennstoffzelle, (www.handelsblatt.com,
         22. November 2019)

[03]   Brian Edgecombe, Next Generation Hydrogen Storage Vessels Enabled by Carbon Fiber Infusion with a Low Viscosity,
         High Toughness Resin System, Materia, Inc. June 10, 2015 DOE AMR, Project ID: ST114

[04]   Scott McWhorther, Grace Ordaz, Onboard Type IV Compressed Hydrogen Storage Systems: Current Performance and
         Cost, DOE Fuel Cell Technologies Office Record, June 11, 2013, US Department of Energy

[05]   Sven Geitmann, Wasserstoff und Brennstoffzellen: Die Technik von morgen. Hydrogeit Verlag, Kremmen 2004,
         ISBN 3-937863-04-4

[06]   Sven Geitmann, Wasserstoffautos: Was uns in Zukunft bewegt. Hydrogeit Verlag, Kremmen Mai 2006,
         ISBN 978-3-937863-30-6

[07]   Dirk Asendorpf, Emissionsfreie E-Autos gibt es gar nicht. Zeit Online. 4. November 2017 (zeit.de,
         16.10.2019)

[08]   Electric vehicles from life cycle and circular economy perspectives, TERM 2018: Transport and Environment Reporting
         Mechanism (TERM) report, European Environment Agency, ISSN 1977-8449

[09]   Kaiser, S., Donnerhack, S., Lundbladh, A., Seitz, A., "A Composite Cycle Engine Concept with Hecto-Pressure Ratio," 51st
         AIAA/SAE/ASEE Joint Propulsion Conference, Orlando, FL, 2015

[10]   Richard van Basshuysen, Fred Schäfer: Handbuch Verbrennungsmotor, Kap. 3. Friedrich Vieweg & Sohn Verlag/GWV
         Fachverlage GmbH, Wiesbaden 2005, ISBN 3-528-23933-6

[11]   Helmut Tschöke, Hanns-Erhard Heinze, Einige unkonventionellen Betrachtungen zum Kraftstoffverbrauch von PKW,
         Magdeburger Wissenschaftsjournal 1-2/2001

[12]   Klaus D. Timmerhaus, Thomas M. Flynn: Cryogenic Process Engineering. Springer Science+Business Media, LLC,
         New York 1989, ISBN 978-1-4684-8758-9

[13]   Sarah Hamdy, Francisco Moser, Tatiana Morosuk, George Tsatsaronis, Exergy-Based and Economic Evaluation of
         Liquefaction Processes for Cryogenics Energy Storage, Creative Commons Attribution (CC BY)

[14]   Bernd Ameel, Christophe T'Joen, Kathleen De Kerpel, P. De Jaeger, Thermodynamic analysis of energy storage with a
         liquid air Rankine cycle, Applied Thermal Engineering, April 2013

[15]   G. Paniaguan, M.C. Iorio, N. Vinha, J. Sousa: Design and analysis of pioneering high supersonic axial turbines.
         International Journal of Mechanical Sciences, von Karman Institute for Fluid Dynamics Chaussée de Waterloo 72,
         1640 Rhode-Saint-Genèse, Belgium

[16]   Dipl.-Ing. Ole W. Willers, Dipl.-Ing. Harald Kunte, Prof. Dr.-Ing. Jörg Seume, ORC-Turbinen-Generatoreinheit für Lkw-
         Anwendungen, ATZ - Automobiltechnische Zeitschrift 10/2017

[17]   Verneau, A, 1987, Supersonic Turbines for Organic Fluid Rankine Cycles from 3 to 1300 kW, Verdonk, G., and Dufournet,
         T., Development of a Supersonic Steam Turbine with a Single Stage Pressure Ratio of 200 for Generator and Mechanical
         Drive, von Karman Institute Lecture Series on "Small High Pressure Ratio Turbines", June.

[18]   Weiß Andreas P., Hauer Josef, Popp Tobias, Preissinger Markus, Experimental Investigation of a Supersonic Micro
         Turbine Running With Hexamethyldisiloxane, 36th Meeting of Departments of Fluid Mechanics and Thermodynamics /
         16th conference on Power System Engineering, Thermodynamics & Fluid Flow - PSE 2017, June 13 - 15, 2017, Pilsen,
         Czech Republic (www.oth-aw.de, 26.11.2019)

[19]   L. Shao, J. Zhu, X. Meng, X. Wei, X. Ma, Experimental study of an organic Rankine cycle system with radial inflow
         turbine and R123, Applied Thermal Engineering. 124 (2017) 940–947. doi:10.1016/j.applthermaleng.2017.06.042.

[20]   Yukihiro Shinohara, Katsuhiko Takeuchi, Dr. Olaf Erik Herrmann, Dr. Hermann Josef Laumen Common-Rail-
         Einspritzsystem mit 3000bar, Motortechnische Zeitschrift 01/2011

[21]   George P. Sutton, Oscar Biblarz, Rocket Propulsion Elements, 7th-ed., Chapter 9 'Combustion of Liquid Propellants', A
         Wiley-Interscience publication 2001, ISBN 0-471-32642-9

[22]   Fei Xing, Arvind Kumar, Yue Huang, Shining Chan, Can Ruan, Sai Gu, Xiaolei Fan, Flameless combustion with liquid fuel: A review focusing on fundamentals and gas turbine application, Applied Energy 193 (2017) 28–51

[23]   Marco Piumetti, Samir Bensaid, Debora Fino & Nunzio Russo (2015) Catalysis in Diesel engine NOx aftertreatment: a review, Catalysis, Structure & Reactivity, 1:4, 155-173, DOI: 10.1080/2055074X.2015.1105615

[24]   Diletta Sciti, Laura Silvestroni, Frédéric Monteverde, Antonio Vinci & LucaZoli, Introduction to H2020 project C3HARME: nextgeneration ceramic composites for combustionharsh environment and space, 117:sup1, s70-s75, DOI:10.1080/17436753.2018.1509822

[25]   Michael C. Halbig, and Mrityunjay Singh, Additive Manufacturing of SiC-Based Ceramics and Ceramic Matrix Composites, NASA Glenn Research Center, Cleveland, OH, Ohio Aerospace Institute, Cleveland, OH

[26]   https://www.sycotec.eu/produkte/industrielle-antriebstechnik/motorkomponenten/turbogeneratoren/, 22.01.2020

[27]   https://www.formula1.com/en/championship/inside-f1/rules-regs/2018-season-changes.html, 04.12.2019

[28]   https://www.racecar-engineering.com/tech-explained/how-to-design-a-motorsport-battery-in-7-steps/, 04.12.2019

[29]   Sven Geitmann, Wasserstoffautos: Was uns in Zukunft bewegt. Kap. 5.2 Kryogen-Behälter, Hydrogeit Verlag, Kremmen Mai 2006, ISBN 978-3-937863-30-6

[30]   Cost and Price Metrics for Automotive Lithium-Ion Batteries. In: www.*energy.gov.* Februar 2017, 01.08.2017

[31]   Sven Geitmann, Wasserstoff und Brennstoffzellen: Die Technik von morgen. Kap. 5.2 Kryogen-Behälter, Hydrogeit Verlag, Kremmen 2004, ISBN 3-937863-04-4

[32]   https://www.stirlingcryogenics.eu/files/__documents/1/StirLAIR-4TechnicalSpecification.pdf, 09.12.2019

[33]   Gnann, T.; Wietschel, M.; Kühn, A.; Thielmann, A.; Sauer, A; Plötz, P.; Moll, C.; Stütz, S.; Schellert, M.; Rüdiger, D.; Waß-muth, V.; Paufler-Mann, D.: „Brennstoffzellen-Lkw: kritische Entwicklungshemmnisse, Forschungsbedarf und Marktpotenti-al", Studie im Rahmen der wissenschaftlichen Beratung des BMVI zur Mobilitäts-und Kraftstoffstrategie der Bundesregie-rung, Fraunhofer ISI, Karlsruhe, Fraunhofer IML, Dortmund, PTV Transport Consult GmbH, Karlsruhe, 2017, www.bmwi.de, 11.02.2020

[34]   https://www.umweltbundesamt.de/themen/co2-emissionen-pro-kilowattstunde-strom-sinken, 11.02.2020

[35]   https://www.autobild.de/artikel/auto-bild-exklusiv-strom-teurer-als-benzin-11360715.html, 11.02.2020

[36]   https://www.ingenieur.de/technik/forschung/heisst-die-loesung-fuer-das-treibstoffproblem-lohc/, 11.02.2020

[37]   https://www.epochtimes.de/politik/deutschland/ab-2021-sie-laden-ab-sofort-mit-reduzierter-stromstaerke-strom-rationierung-fuer-private-e-autos-a3078343.html, 11.02.2020

[38]   Adriano Sciacovellia, Daniel Smitha, Helena Navarroa , Yongliang Lia, Yulong Ding, Liquid air energy storage – Operation and performance of the first pilot plant in the world, Birmingham Centre for Energy Storage, School of Chemical Engineering, University of Birmingham, UK June 2016

YOUR KNOWLEDGE HAS VALUE

- We will publish your bachelor's and
  master's thesis, essays and papers

- Your own eBook and book -
  sold worldwide in all relevant shops

- Earn money with each sale

Upload your text at www.GRIN.com
and publish for free